Move It!

by Jamie A. Schroeder

Table of Contents

How Do Objects Move? 4
What Objects Move Back and Forth? 10
What Objects Spin? 16
Glossary and Index 20

I need to know these words.

axis

gravity

motion

pendulum

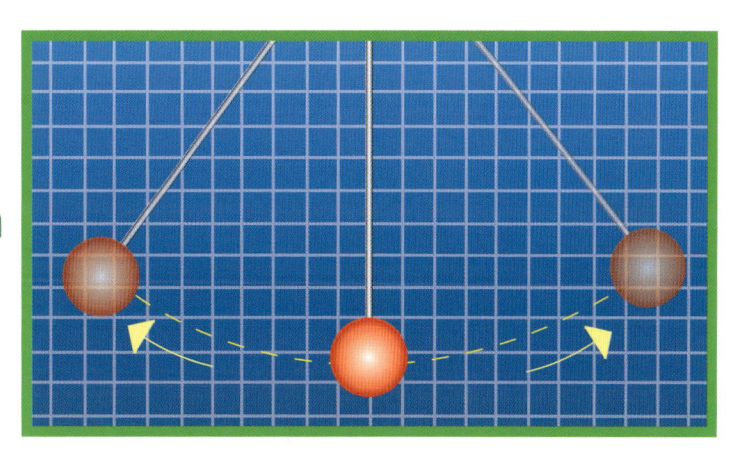

pull

push

How Do Objects Move?

Look at the baseball. The man hits the baseball with a bat. The baseball moves fast. The baseball travels through the air. The baseball is in motion.

▲ The bat sends the ball into the air. The ball is in motion.

The soccer player kicks the ball. The kick is a **force**. The force pushes the ball. The ball is in motion. The force causes the motion.

▲ The player kicks the ball. The ball is in motion.

5

Some forces are strong. A strong force can move an object far. Some forces are weak. A weak force might not move an object very far.

▲ A kick is a strong force. A tap is a weak force.

Do you play tug-of-war? You pull a rope. You use a force when you pull. The rope moves. The rope is in motion. You are in motion, too.

pull

▲ The children pull the rope. The pull is a force. The force moves the children.

7

Most **surfaces** are not smooth. A surface might be rough. A surface might be bumpy. The surfaces rub against moving objects. The rubbing is **friction**. The friction slows moving objects.

Try This

Look at the pictures. Each picture shows a different surface. Match each picture with one of the words.

smooth
rough

A batter hits a ball. The ball rolls on the ground. Then the ball stops. What makes the ball stop moving?

▲ The ground causes friction. The friction makes the ball slow down.

What Objects Move Back and Forth?

Think of a swing. A swing moves back and forth. A swing is a **pendulum**. A pendulum is a weight on a rope. The pendulum is the end.

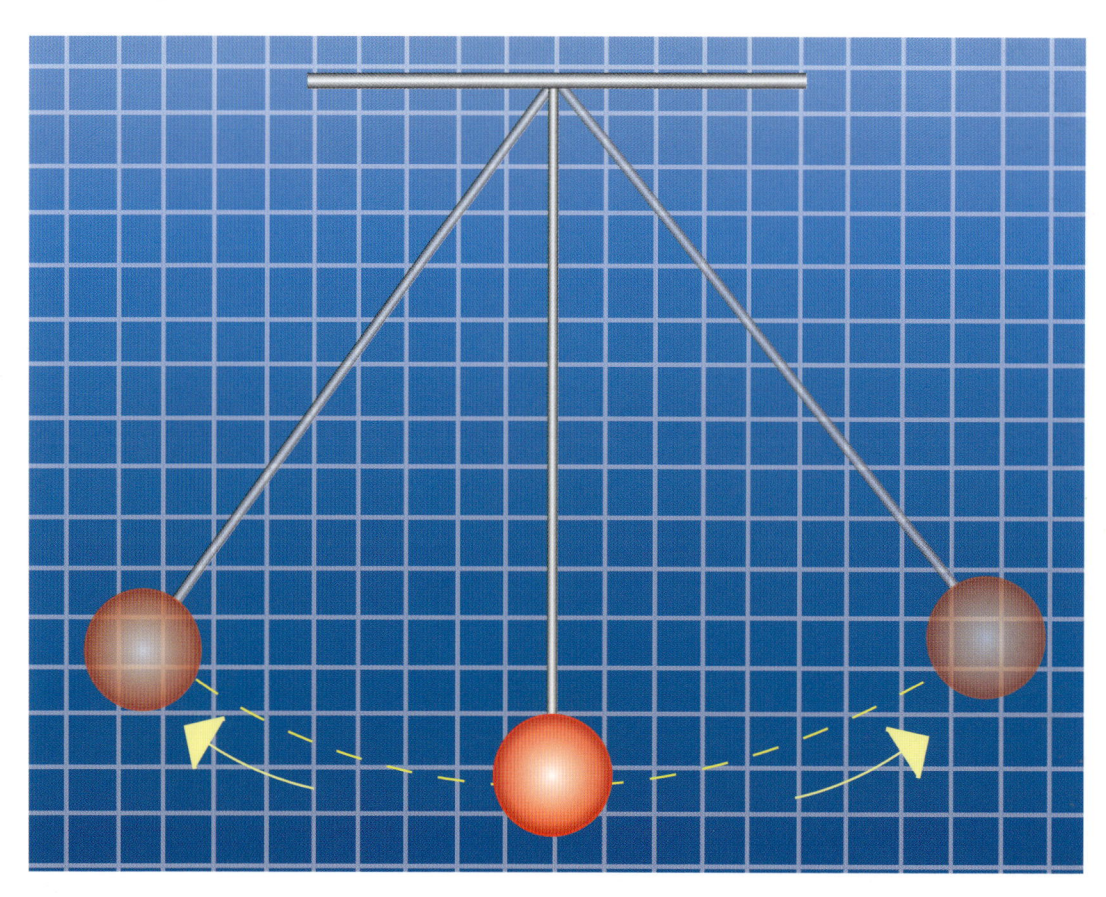

▲ A pendulum moves back and forth.

A wrecking ball is a pendulum. Some clocks have a pendulum. Can you think of another pendulum?

▲ A pendulum can help keep time.
A pendulum can help knock down buildings!

A swing in a park moves. You push the swing. The swing moves. You stop pushing. Why does the swing stop moving?

▲ The swing moves back and forth.

Gravity is pulling on the swing. Gravity is always pulling on objects.

▲ The swing stops moving because of gravity.

13

Have you ever dropped some food? The food falls because of gravity. Why do objects fall? Gravity pulls objects toward Earth.

▲ Objects fall because of gravity.

14

People stay on the ground. A force pulls on the people. The force is gravity. People would float in the air without gravity!

Did You Know?

Gravity is strong on Earth. Gravity is weaker on the moon. You weigh sixty pounds on Earth. You would weigh about ten pounds on the moon!

What Objects Spin?

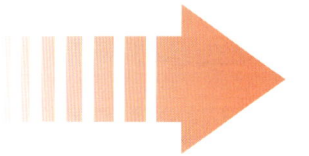

Look at the girl. The girl spins the top. The top spins on an **axis**. The top moves in circles.

▼ The top spins. The top is in motion.

An axis is a line. An object spins around the line. Look at the axis in the picture. What other things spin around an axis?

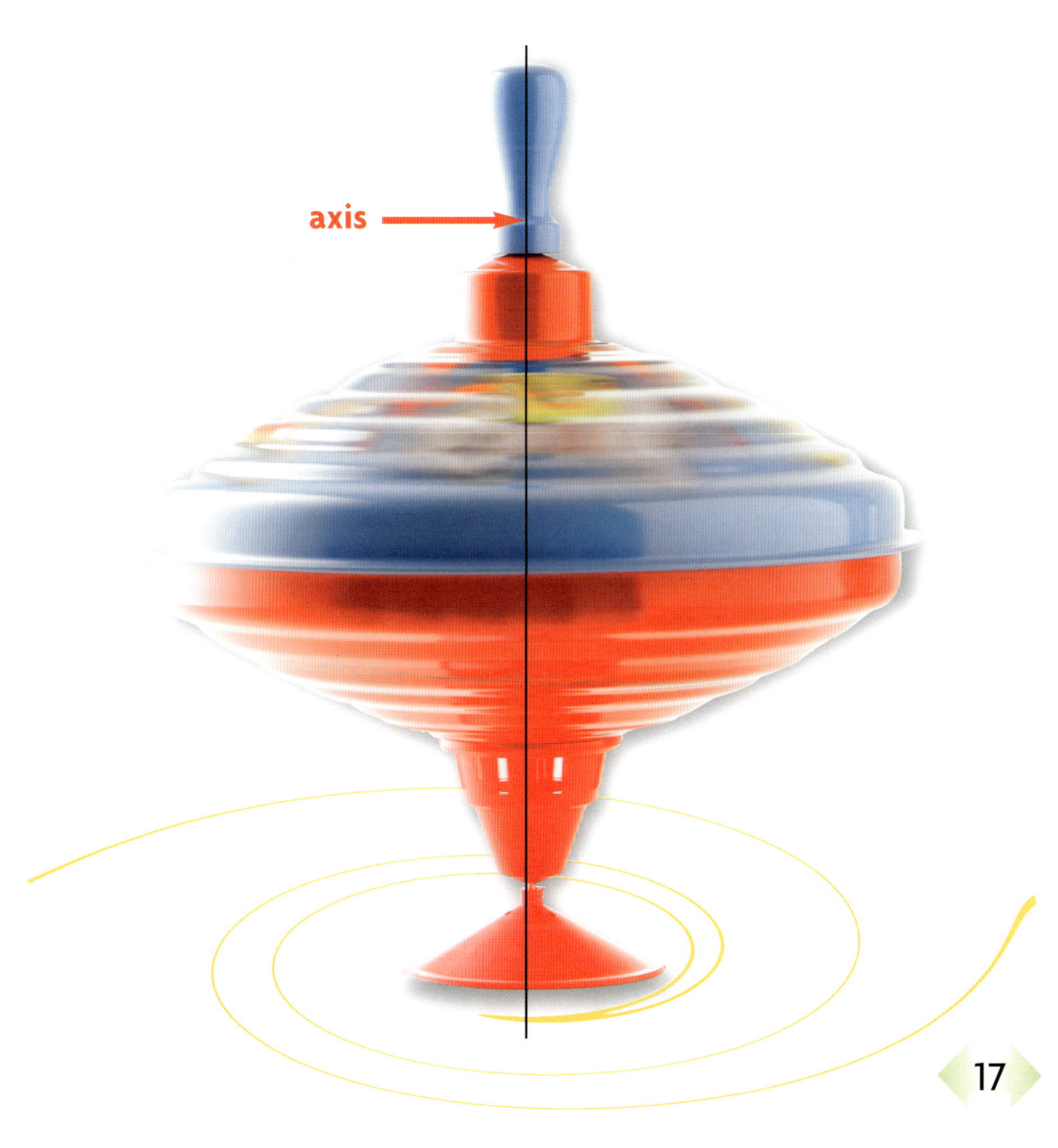

axis

Earth rotates, or spins, on an axis.
Earth rotates all the time.

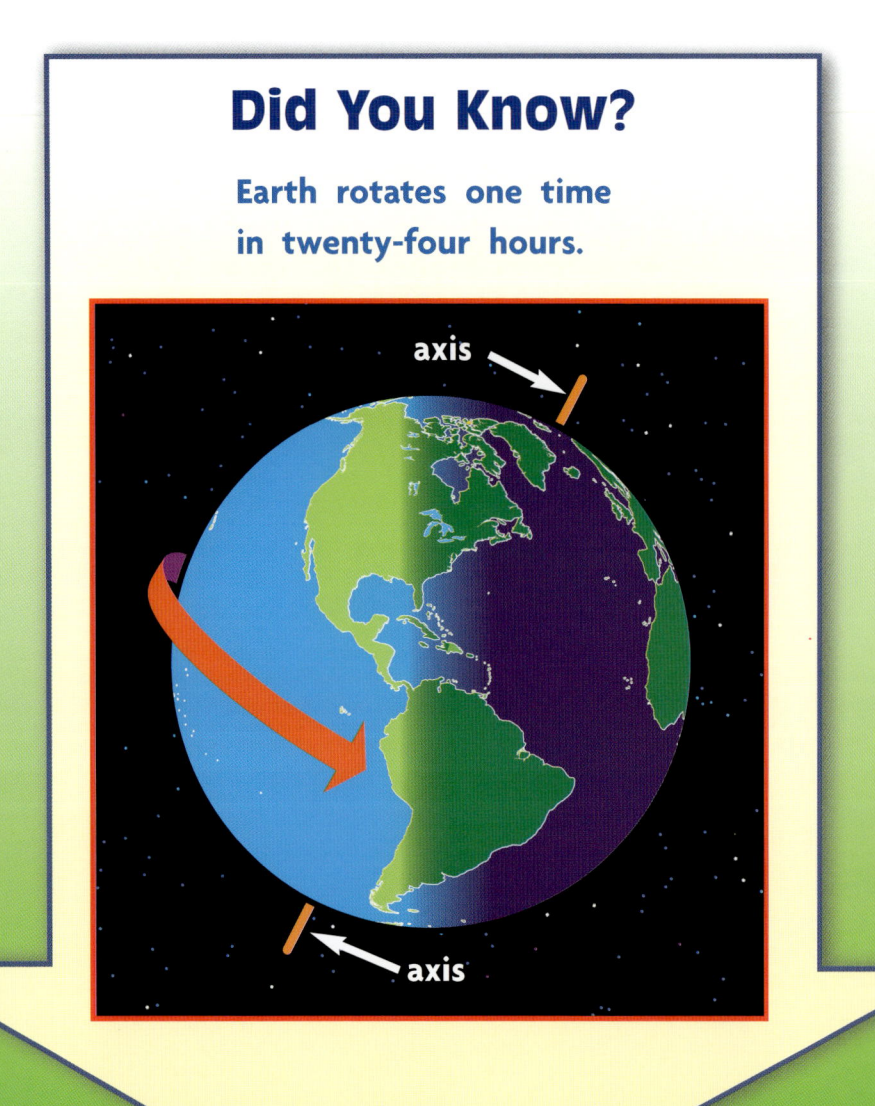

Did You Know?

Earth rotates one time in twenty-four hours.

axis

axis

Look at the objects around you. Think about how the objects move. Do you push the objects? Do you pull the objects? What direction do the objects move in?

▲ What causes these objects to move?

Glossary

axis (AK-sus): a straight line that an object turns around
See page 16.

force (FORS): any cause that makes an object move or stop
See page 5.

friction (FRIK-shun): a force that happens when one object rubs against another object
See page 8.

gravity (GRA-vuh-tee): the force that pulls objects
See page 13.

pendulum (PEN-juh-lum): a weight or object that moves back and forth from a point
See page 10.

surface (SER-fuss): the outside or top parts of an object
See page 8.

Index

axis, 16–18
direction, 19
Earth, 14–15, 18

force, 5–7, 15
friction, 8
gravity, 13–15

pendulum, 10–11
surfaces, 8